I0473344

THIS NOTEBOOK

BELONGS TO

www.ingramcontent.com/pod-product-compliance
Lightning Source LLC
Chambersburg PA
CBHW081003170526
45158CB00010B/2895
* 9 7 8 1 0 7 2 5 7 4 7 4 3 *